YOUR KNOWLEDGE HAS VALUE

- We will publish your bachelor's and master's thesis, essays and papers

- Your own eBook and book - sold worldwide in all relevant shops

- Earn money with each sale

Upload your text at www.GRIN.com and publish for free

Bibliographic information published by the German National Library:

The German National Library lists this publication in the National Bibliography;
detailed bibliographic data are available on the Internet at http://dnb.dnb.de .

Imprint:

Copyright © 2018 GRIN Verlag
Print and binding: Books on Demand GmbH, Norderstedt Germany
ISBN: 9783668668294

This book at GRIN:

https://www.grin.com/document/416675

Ronald Edilberto Ona

Community Based Sustainable Tourism in Puerto Princesa City. Ugong Rock Adventures Case History 2014

GRIN Verlag

Ugong Rock Caving, Spelunking and Zipline Adventure Community-based Sustainable Tourism:

An Organizational Case History

Table of Contents

List of Tables

List of Figures

Acronyms and Abbreviations

UPM	University of the Philippines Mountaineers
ELAC	Environmental Legal Assistance Center
TCTA	Tagabinet Community Tourism Association
UNDP	United Nations Development Program
PPUR	Puerto Princesa Underground River
TURSCO	Tagabinet Ugong Rock Tourism Service Cooperative
AFI	ABS-CBN Foundation Incorporated
CBST	community-based sustainable tourism
PhP	Philippine Peso
URA	Ugong Rock Adventures
PSU	Palawan State University
FGD	Focus Group Discussion
KII	Key Informant Interviews
HHI	Household Interviews
CWA	Charity Women's Association
CTR	Community Tourism Representative
SEC	Security and Exchange Commission
ATV	All-Terrain Vehicle
ECC	Environmental Compliance Certificate
LGU	Local Government Unit
CDF	Community Development Funds
MOA	Memorandum of Agreement
DOT	Department of Tourism

CHAPTER 1. INTRODUCTION

Bunglon Cave is a jagged karst limestone tower standing out of the flatlands of alluvium and soil in the vicinity of Barangay Tagabinet. According to anecdotes from senior residents of the area, the name Bunglon is a Tagbanua term for wild dogs that used to roam the cave. The Cave is geographically situated at North 11°54'71" and East 70°31'73". From the 1970s until the 1990s, the hill was trekked by young people for its scenic and natural beauty and became a popular hangout for the locals. Aside from its recreational service, Bunglon's forest provided an important livelihood for the Tagabinet community whose occupations were mainly farming and fishing. During that period, slash and burn or *kaingin* farming was prevalent as well as the conversion of mangrove forests to fish pens. The highly extractive and destructive nature of these two practices took its toll on the natural environment of Bunglon which in turn seriously affected the livelihood of the community. In 1999, Mr. Kenneth Kennedy, and some member of the University of the Philippines Mountaineers (UPM) discovered one particular stalactite inside the cave which produced a haunting humming sound when knocked. He named it "Ugong Rock" from the Tagalog word *"ugong"* for the humming sound it made. Since then, Bunglon Cave became more popular among the community members and visitors as Ugong Rock.

In the year 2000, the Environmental Legal Assistance Center (ELAC), an environmental nongovernment organization based in Puerto Princesa City visited Ugong Rock to study the unabated environmental degradation happening in the area as part of their project proposal on community-based ecotourism in the Ulugan Bay area. ELAC provided technical and legal assistance to discourage people from their environmentally destructive practices and to stop illegal activities conducted by some individuals who acted purely on self-interest and disregarded the negative impacts of their activities to the environment. As part of ELAC's activities on environmental awareness in the Ulugan Bay area, they organized the Tagabinet Community Tourism Association (TCTA) whose members were Barangay Tagabinet residents who decided to join the proposed community-based tourism project of ELAC.

Once the illegal activities were halted, a new problem emerged for the Tagabinet community – the problem of alternative livelihood. Since most of them do not wish to continue with slash and burn farming and cutting down the mangroves, they had to find an alternative livelihood which they can rely on for a long time. Finding employment in the city was a difficult choice given the low level of educational attainment of many of the community members. Fortunately, ELAC's project proposal which was submitted to the United Nations Development Program (UNDP) was approved and a UNDP Grant of about PhP150,000.00 was given to TCTA to put up an Information Center which served as a receiving and orientation area for visitors. Because of Ugong Rock's accessibility to Sabang and the Puerto Princesa Underground River (PPUR), local and foreign tourists soon discovered the beauty of the place. Travel and tour agencies recognized the potential of Ugong Rock as a complementary tourist attraction to the PPUR. From its initial form as the Tagabinet Community Tourism Association (TCTA) in 1999 to its current structure as the Tagabinet Ugong Rock Tourism Service Cooperative (TURSCO), the ABS-CBN Foundation Incorporated (AFI) in cooperation with the Puerto Princesa City Tourism Office played central roles in what can be called as the first real successful community-based sustainable tourism (CBST) project in Palawan. As a gauge of its success, visitors (local and foreign tourists) steadily increased from only 109 in 2008 to 31,056 visitors in 2011 translating to a dramatic increase in income from PhP 7,150.00 to PhP 7,549,495.35. For 2012, records from TURSCO showed a total of 47,183 visitors translating to an annual income of PhP 13,843,146. Latest data for 2013 showed visitor arrivals reaching a record high of 50,385.

Because of the lack of a process documentation research (PDR) or an organizational case history for the TCTA/TURSCO Ugong Rock CBST, the dynamic processes, social relationships, challenges, conflicts and best practices that happened as part of the Ugong Rock Adventures (URA) success story could end up as footnotes in its still unfolding story. For this reason, this organizational case history is written in the hope that it could serve as a useful template for funding agencies and local communities in the Philippines or elsewhere with ecotourism potentials that aspire to duplicate the TCTA/TURSCO success story. This organizational case history is one of the two studies conducted by the Palawan State University (PSU) for the URA CBST Project. The other study focused on the assessment of socioeconomic and environmental impacts of URA to Barangay Tagabinet.

CHAPTER 2. METHODOLOGY

The method used for this study is the review and analysis of organization documents from TCTA/TURSCO. Other available secondary data were also gathered and examined. Data were validated through a Focus Group Discussion (FGD) with selected members and officers of TURSCO and Key Informant Interviews (KIIs) with AFI coordinators for the URA CBST. Additional informations were obtained from Household Interviews (HHIs) conducted as part of the socioeconomic impact assessment of the URA CBST by the Palawan State University. In terms of the analysis and interpretation of data, this study focused on four subject areas: 1) Organizational Context: history, physical and social settings; 2) Strategies Described: approaches taken, actors involved, and institutional support; 3) Challenges: concerns and problems that emerged and various perspectives; and 4) Outcomes: accomplishments, changes, lessons learned and best practices.

CHAPTER 3. RESULTS AND DISCUSSION

3.1   Organizational Context

3.1.1   Physical and Social Settings

Ugong Rock is located within Barangay Tagabinet which is found in the northeastern flank of Ulugan Bay, along the northern section of the road from Puerto Princesa City to Sabang. It includes a short stretch of the Ulugan Bay coastline featuring mangrove, sea grass and coral ecosystems. The main settlement area however is situated in the Tagnipa Inlet (Socrates 2003).

The Ugong Rock karst formation is both an ecotourism site and community within Barangay Tagabinet, Puerto Princesa City (Figures 1 and 2). URA is ideally placed at the gateway to the PPUR and serves as a model exposure of the St. Paul Limestone Formation for demonstrating the features, not only of the karst but also of the unique geology of Palawan, to recreational and educational tourists. The Ugong Rock Cave and Zipline Adventures is the CBST project owned and operated by TURSCO with technical and management support from AFI.

Removed due to copyright reasons - the editor

Figure 1. ECAN map of Ugong Rock in Barangay Tagabinet

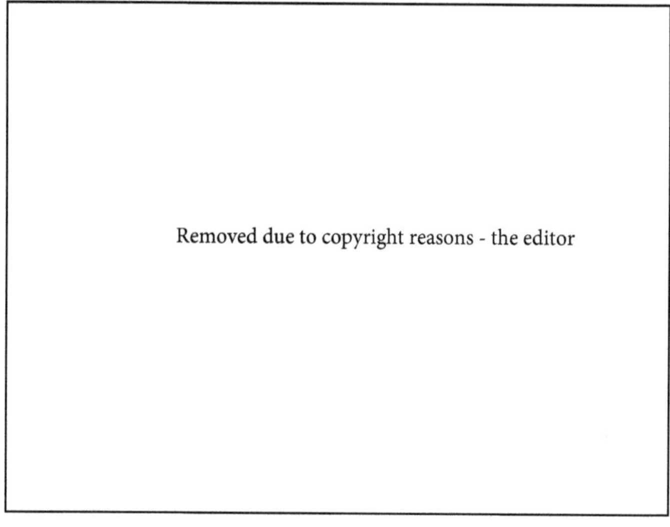

Figure 2. The Ugong Rock karst formation in Barangay Tagabinet

3.1.2   Timeline

Timelines are graphical representations of history drawn by members of an organization or community based on their personal recollections of events starting from the time they considered as the "starting point" to recent events prior to the making of the timeline. Based on the FGD conducted on 20 January 2014 at the TURSCO office, members and officers narrated the organization's history and drew a timeline of the URA CBST project (Table 1). The timeline matrix produced covered the period from 1992 to 2014.

Table 1. Timeline matrix showing the history of Ugong Rock Cave and Zipline Adventures (Source: TURSCO)

Year	Event/Activity
1992	• Former Mayor Edward S. Hagedorn ordered the development of Ulugan Bay Area as a potential tourist destination. One of the sites identified was Ugong Rock.
1998	• Charity Women's Association (CWA) was formed with the initiative of Ms. Girlie Bayron of the City Tourism Office. The members are the women senior citizens of Barangay Tagabinet. Their activities include cleaning and maintaining the area surrounding Ugong Rock.
1999	• Martin L. Felstead, a consultant from the United Nations Development Program (UNDP), visited the area in connection the development of the proposal for the

Year	Event/Activity
	Community-Based Eco-Tourism in Ulugan Bay. • Kenneth Kennedy and the UP Mountaineers visited the area; the name was changed to Ugong Rock because of the sound that the caves produced. • Environmental Legal Assistance Center (ELAC) and the City Government organized Tagabinet Community Tourism Association (TCTA). The association members were barangay residents who wanted to be involved in the Ugong Rock proposal. • A UNDP Grant worth PhP 150, 000.00 was given to Barangay Tagabinet. The money was used to develop the Information Center, a receiving and orientation area for visitors.
2000	• UNDP Consultant Martin L. Felstead submitted the Master Plan for Community-Based Eco-Tourism in Ulugan Bay, part of the proposal is the development of CBST. One of the activities proposed was to develop spelunking activities in Barangay Tagabinet. • Dr. Jose Antonio U. Socrates did the baseline data and geological assessment of Ugong Rock. He also helped in developing the script for guides during visitor orientations. • Former Mayor Edward S. Hagedorn endorsed the proposed establishment of Ugong Rock Spelunking and Summit View Deck of barangay Tagabinet. The project aims to provide supplemental budget to the residents of the barangay by offering a natural geological formation as a tourist attraction and to integrate local livelihood development needs with those of environmental conservation.[1] • A mining application in Barangay Tagabinet was prevented by ELAC and TCTA.
2001	• TCTA elected its first set of officers.
2002	• TCTA and ELAC demolished the illegally constructed *bangus* fishponds.
2003	• ELAC registered TCTA to the Security and Exchange Commission (SEC). • An application for Environmental Compliance Certificate was submitted. • A UNDP Grant amounting to PhP 37, 000.00 was given to TCTA to support livelihood projects (sari-sari store, micro-lending, hog/swine raising, etc.)
2004-2007	• Tourism was adversely affected by the Dos Palmas kidnapping incident. • Some members went back to their former livelihood (charcoal making, *kaingin*, etc.) to support their family. • The organization still conducted regular meetings despite the setbacks.
2008	• Dr. Gerardo "Gerry" V. Ortega came to Barangay Tagabinet and reorganized TCTA. • TCTA's mission and vision were evaluated and new members were recruited. • AFI issued a grant worth PhP 100, 000.00 for TCTA. The money was used to

---

[1] Initial Environmental Examination, City Government of Puerto Princesa

Year	Event/Activity
	construct the Information Center and repair the facilities.
	• URA or Ugong Rock Spelunking and Summit View Deck was formally launched.
2009	• Caving was introduced as part of the activities for tourists.
	• The site became an alternative/supplemental route for tourists going to Underground River, thus ensuring the inflow of tourists.
2010	• The construction of a zip line begun. The zip line was adapted by TCTA, as suggested by their guests.
	• The City government gave PhP 800,000.00 for the construction of a zip line.
	• A soft launch of the zip line was held.
2011	• The 330 meter zip line was formally launched.
	• A proposal for the addition of a 400 meter two-rider zip line was submitted.
2012	• Soft launch of superman zip line, rappelling, and rock-climbing activities.
	• The URA closed for 40 days because of a zip line accident.
2013	• The Superman zip line was formally launched.
	• The souvenir shop, spa, bookstore, restaurant and fruit shake stand were also opened.
	• TCTA became a cooperative and the name was changed to TURSCO.
2014	• Future plans of TURSCO include creating a financial plan, a savings account for every member, the addition of kayaking activities, and developing an alternative livelihood (farming) for members during the lean months for tourism.

Based on the timeline matrix, Ugong Rock CBST Project started out in 1999 from the idea of establishing a CBST project in the Ulugan Bay area as recommended in a project proposal that ELAC submitted to UNDP. The UNDP awarded a grant of PhP 150,000.00 to Barangay Tagabinet for the construction of a visitor's information center. As part of the project, TCTA was organized by ELAC as a core group/community who will manage the Ugong Rock CBST. In 2003, TCTA had only 19 volunteer pioneer members with an initial fund of PhP 37,000.00 from UNDP as TCTA's share from the PhP 150,000.00 general fund for the CBST project in Ulugan Bay. It was also during this year that TCTA became a Securities and Exchange Commission (SEC) registered association. Its Environmental Compliance Certificate (ECC) application was also submitted.

By 2008, AFI-Palawan headed by the late Dr. Gerry V. Ortega, became actively engaged with TCTA resulting to its reorganization, evaluation of its TCTA Vision – Mission statement and the recruitment of new members. As part of AFI's involvement, a grant of PhP 100,000.00 was given to TCTA for the construction of a tourist's information center. In 2010, the Puerto Princesa

City government funded the construction of zip lines in Ugong Rock to the cost of PhP 800,000.00.

Since then, the TCTA has metamorphosed into a service cooperative as TURSCO with 10 officers (board of directors) and 35 members. It is not surprising that URA is becoming the most economically successful CBST project in Puerto Princesa City based on the latest annual income and total number of visitors recorded.

3.2   Strategies Described

3.2.1   Approaches Taken

3.2.1.1 1998-1999: Top – Down Approach

Some members of the TURSCO traced their early beginnings to the Charity Women's Association (CWA) organized by Mrs. Girlie Bayron of the Puerto Princesa City Tourism Office in 1998. As an all–women's group, the CWA was mainly involved in clean–up and beautification activities in the Ugong Rock area. Basically, the approach was "cosmetic" with beautification and cleanliness as underlying agenda in the promotion of tourism in the area. During that time, Ugong Rock was not considered as tourist spot but was just a part of the larger Ulugan Bay potential tourism region. In terms of organization, CWA can be considered as a "top–down" group where authority and command is mainly from the top (City Tourism Office) directed towards the members. Active community involvement was absent as membership was limited only to selected women members and CWA had no agenda of its own.

3.2.1.2 2000-2003: The Community Based Sustainable Tourism Approach

When the Tagabinet Community Tourism Association was formed in 1999, it was an organization of volunteers willing to become part of a proposed CBST project in Barangay Tagabinet. The TCTA was formed upon the initiation of ELAC which was looking for local residents who will volunteer to prepare Ugong Rock as an ecotourism site. It is important to note that at the inception stage of the TCTA, there was no operational ecotourism product yet and commitments to join the association entailed a lot of risks and uncertainties. At this stage, the potential of Ugong Rock as an ecotourism site was already recognized by ELAC and the Puerto Princesa City government. The Ugong Rock Spelunking and Summit View Deck project was proposed to provide supplemental livelihood to the residents of the barangay. By offering Ugong Rock as a tourist attraction, local livelihood development needs are integrated with environmental conservation.

In terms of operation, ELAC was responsible for providing organizational and management assistance as well as financial support. While TCTA had elected officers, both officers and members looked up to ELAC as an overseer since they considered themselves as still lacking in organizational and management skills. Although the authority structure of TCTA resembled a top-down structure, the degree of authority exerted by ELAC was minimal since there was a distinct organizational framework (constitution, mission-vision etc.) that demarcated the roles and functions, as well as rules that governed the power relationship between TCTA and ELAC. By 2003, the operation of Ugong Rock as a tourist site began. In the same year, a grant of PhP 37,000.00 from UNDP was given to TCTA for support livelihood projects (sari-sari store, micro-lending, hog raising etc.).

3.2.1.3 2004-2007: Period of Decline and Inactivity

The period 2004-2007 was marked by a decline in tourist arrivals in Palawan due to the high-profile kidnapping incident involving Abu Sayyaf terrorist group and foreign tourists in Dos Palmas resort in Honda Bay. Because of the negative publicity caused by the kidnapping, income from the tourism sector drastically declined and severely affected the income of the TCTA. With the decline of tourism in Palawan and the loss of income for TCTA members, some members "resigned" but most members reverted back to their pre-Ugong Rock livelihoods which were mainly extractive in nature such as charcoal making, rattan gathering, and slash and burn farming much to the detriment of the environment. While Ugong Rock tourism took a major dip in terms of visitors and income, the TCTA still managed to hold regular meetings. The approach used by TCTA seemed to be "wait and see" as typical Filipinos would sometimes cite "*bahala na*" or come what may attitude believing that the tourism slump and negative publicities will eventually pass.

3.2.1.4 2008-2011: Period of Renewal and Redirection

By 2008, Palawan was slowly recovering from the tourism slump and bad publicity caused by the Dos Palmas kidnapping incident. The Puerto Princesa City Government, in partnership with ABS-CBN Corporation through AFI, took interest in restarting the Ugong Rock CBST project. Though community interest in the Ugong Rock CBST was its lowest during this period, most TCTA members agreed to give the Ugong Rock CBST project a second chance.

AFI founded Bantay Kalikasan Palawan and recruited the veterinarian and environmental activist Dr. Gerardo V. Ortega to spearhead the project. Mainly through the efforts and personal dedication of the late Dr. Gerardo "Gerry" V. Ortega and Bantay Kalikasan community organizer Marlon Tamsi, the reorganization, the re-evaluation of Vision-Mission Statement, and the recruitment of new members were initiated. To back up this infusion of ideas and leadership, AFI awarded a grant of PhP 100,000.00 to TCTA which was used to construct the Tourist Information Center and repair existing facilities which suffered neglect due to lack of funds.

During this period, URA was finally launched. By 2009, caving was introduced as part of the recreational activities that tourists can enjoy. This resulted to an increase in visitor arrivals which meant financial returns for TCTA. The approach used during the renewal and redirection phase can be described as the mentor-mentee wherein the mentor AFI through Dr. Ortega and Mr. Tamsi personally guided the TCTA by demonstrating leadership by example and direct participation in TCTA's activities.

3.2.1.5 2009-2010: Bottom-up and Autonomy

The period 2009-2010 was marked by a steady increase in visitor arrivals to Ugong Rock due to the inclusion of Ugong Rock as a "side trip" for tourists going to Sabang for the Underground River tour. This significant increase was brought about by the positive endorsement and cooperation of travel and tours agencies in Puerto Princesa City specifically the shuttle van drivers and their partner tour guides. The Ugong Rock CBST project was able to recover from the economic and psychological slump that occurred during the 2004-2007 period. The recovery of the project had a positive effect on both the leadership and members of the TCTA.

With regular visitors to Ugong Rock, income from tourism became stable and TCTA became financially self-sufficient that more community members from Barangay Tagabinet wanted to join the association and partake in the benefits from Ugong Rock ecotourism. After achieving financial autonomy, the leaders of TCTA became more confident of their business management and organizational skills derived from the years of experience in operating the URA. At this period the leadership and authority of TCTA can be described as "bottom-up" wherein decision making starts from members themselves (bottom) and ends up with the officials of the association (top) through a democratic process of free discussion, consensus, and voting. It was also in 2010 that the construction of zip lines began. This idea came from URA tourist visitors who wanted to experience an adventurous and breathtaking descent through zip line from Ugong Rock. This idea was accepted by TCTA and zip lines soon became a reality.

3.2.1.6 2011-2012: Maturity and Full-autonomy

The start of 2011 was a bleak and sad one for TCTA and the Ugong Rock CBST Project due to the untimely death of Dr. Gerry Ortega who was recognized by TCTA and Tagabinet community members as their mentor and "leader". The sudden death of Dr. Ortega was strongly felt by the community who considered him as a "father" and some kind of spiritual leader.

Ordinary associations with no or limited interpersonal relationship and mutual positive reinforcement will suffer a major setback when a dominant leader passes away. However due to TCTA's experience and training in bottom-up approach of decision making and organizational management, as well as the dynamic mentorship provided by the late Dr. Ortega, TCTA managed to survive the crisis. At this point, it can be said that TCTA as a community organization has matured and has become fully autonomous.

3.2.1.7 2013-Present: Going Cooperative and Becoming Mainstream

In order to maximize income and community benefits as well as to enhance financial viability, TCTA decided to become a full cooperative in 2013. Thus, by 2013 the Tagabinet Community Tourism Association became officially known as the Tagabinet Ugong Rock Service Cooperative (TURSCO). With an annual income posted at PhP 13,843,146.00 for 2012 and PhP 12,717,725.00 for 2013 (AFI 2013), TURSCO's financial viability has been established. Based on TURSCO's financial record from 2008 to 2013, the cooperative earned a total gross earnings of PhP 25,560,871.00 for the past six years making it the most successful community based ecotourism project in Puerto Princesa City. In terms of the future, TURSCO laid out its plans starting 2014. These include: a) financial plan, b) individual savings account for all members, c) kayaking and d) alternative livelihoods for members during tourist lean season. All of these plans suggest that TURSCO-Ugong Rock Cave and Zip Line Adventures is on its way of becoming a mainstream ecotourism provider-product in Puerto Princesa City.

3.2.2   Actors Involved

The following persons have direct or indirect involvement in one way or the other in the creation of TCTA/TURSCO and the development of the Ugong Rock Community-based Sustainable Tourism Project:

1. **Mayor Edward S. Hagedorn** – in 1992, ordered the study and development of Ulugan Bay Area as a potential tourism destinations

2. **Mrs. Girlie Bayron** – through the City Tourism Office in 1998 initiated the formation of CWA composed of elderly women from Barangay Tagabinet, whose main role was for the cleanliness and beautification of Ugong Rock area

3. **Mr. Kenneth Kennedy and the U.P. Mountaineers** – Mr. Kennedy together with U.P. Mountaineers members climbed and explored Ugong Rock cave and discovered that a certain rock formation created humming or "ugong" sound when tapped. The name Ugong Rock became popular and soon was accepted by community members as the new name for Bunglon Cave. Mr. Kennedy formally informed Mr. Felstead, ELAC, City and Barangay Officials about the potential of Ugong Rock as a tourism site.

4. **Martin L. Felstead** – Mr. Fealsted, was a technical consultant of the UNDP who visited the Ulugan Bay area in Puerto Princesa City to conduct inspection and evaluation of the

area in connection with the proposal of the Puerto Princesa City government to set up community-based ecotourism projects in the Ulugan Bay area, he wrote the Community-based Ecotourism Master Plan and the community-based sustainable tourism project of the City Government became operational in 1999

5. **Dr. Jose Antonio U. Socrates** – the late Dr. Socrates did the baseline data and geological survey of Ugong Rock, he also wrote the scripts that Ugong Rock tour guides read during the orientation for guests and tourists

6. **Dr. Gerardo "Gerry" V. Ortega** – the late Dr. Ortega was the Director of the Bantay Kalikasan – Palawan project of AFI, he was the one responsible for the reorientation and reorganization of the TCTA in 2008

7. **Mr. Marlon Tamsi** – Mr. Tamsi was the AFI community organizer who assisted Dr. Ortega in reorganizing and reorienting TCTA

8. **Ms. Jet Millare** – assisted Mr. Tamsi in community organizing work

3.2.3 Institutional Support

The following institutions were involved in different stages of the creation and operation of the Ugong Rock CBST Project:

1. **Puerto Princesa City Government** – the Puerto Princesa City Local Government Unit (LGU) was and is still the primary institution who conceptualized, developed and initiated the creation of CBST Projects in Puerto Princesa City and specifically in the Ulugan Bay area

2. **Environmental Legal Assistance Center** – ELAC organized TCTA and gave it an environmental orientation, as part of its involvement in environmental protection activities, TCTA joined ELAC in dismantling illegal fish pens in the Ulugan Bay area within the boundary of Barangay Tagabinet in 2002 and ELAC registered TCTA with the SEC in 2003.

3. **United Nations Development Program** – UNDP granted Barangay Tagabinet with PhP 150,000.00 for the construction of Information Center for guests and visitors to Ugong Rock in 1999. In 2003 it provided an additional grant of PhP 37,000.00 to TCTA for livelihood support of its members.

4. **ABS-CBN Foundation Inc.** –AFI through its project Bantay Kalikasan Palawan granted TCTA the amount of PhP 100,000.00 for the construction of a visitors information center and receiving hall within the premises of the Ugong Rock CBST. The Information Center is different from the one constructed by Barangay Tagabinet from UNDP funds, the entry of AFI in the Ugong Rock CBST project helped TCTA to recover from the tourism slump that occurred after the Dos Palmas kidnapping incident.

5. **Tagabinet Barangay Council** – The council entered into a Memorandum of Agreement (MOA) on income sharing and Community Development Funds (CDF) to show its support.

6. **Bayan Academy** – Bayan Academy of the ABS-CBN Foundation Inc. provided loans to TURSCO for the construction of Sari-Sari Store and zip line photo souvenir shop

7. **Department of Tourism** – The national government thru the Department of Tourism provided the initial fund for the creation of CBST project in Ugong Rock area before the MoA between TCTA and the Puerto Princesa City Government was finalized

3.3   Challenges

Challenges are problems and concerns that the organization has encountered in the past or is encountering at the present time. Challenges might be internal or external to the organization and is inherent. The way the organization copes up with the challenges and resolve the problems and concerns that face it will determine its future status.

Based on the FGD and KIIs with officers and members of TCTA/TURSCO, the following challenges are presented:

1. **Dos Palmas Kidnapping Incident** – the unexpected kidnapping by the international terrorist group Abu Sayyaf of foreign tourists in Dos Palmas Resort in Honda Bay created a deluge of negative publicities that resulted to the sudden drop of tourism in Palawan, this incident affected the TCTA era of Ugong Rock CBST in 2004-2007 with most members reverting back to their old livelihood involving destructive extraction of natural resources in Ugong Rock area.

2. **Death of Dr. Gerry Ortega** – the sudden death of Dr. Ortega on January 24, 2011 at the time when TCTA was being reorganized and reoriented resulted to a temporary "slowdown" of TCTA activities and affected the morale of members. TCTA members enjoyed a mentor-mentee and close personal relationship with Dr. Ortega thus, his untimely demise was very painful for TCTA members.

3. **Zipline Accident** – the unexpected accident of two foreign tourists in the zip line occurred in 2012. The victims were hospitalized after sustaining major injuries from the accident. Claims for damages were filed and the incident resulted to a 40-day closure of URA. This closure affected the income of the organization and further payments for the damage claims of the two tourists resulted to a substantial debt to AFI. The foundation settled the claims of the tourists in behalf of TCTA but the organization is still paying AFI for the debt.

4. **Sustainability of Ugong Rock CBST** – by 2013, Ugong Rock CBST has established financial viability and organizational stability needed to operate a growing ecotourism venture, however, at the back of their minds members are thinking of the vulnerability of ecotourism to terrorism, pollution and diminishing interest of tourist in Palawan, their main concern is that the high income that they are earning presently might not continue in the future especially with the growing operational expenses that are associated with a growing business.

5. **Conflict with Barangay Officials of Tagabinet** – this can be traced way back in 1999 when UNDP granted Barangay Tagabinet PhP 150,000.00 for the construction of an Information Center which was constructed beside the road going to Sabang instead of inside the vicinity of the URA. The said Information Center is now abandoned and in rundown condition.

6. **Conflict with some non-members in Barangay Tagabinet** – this can be traced in the very beginning of the formation of the TCTA where many barangay residents who did not join the group ostracized TCTA members. When TCTA and Ugong Rock CBST became successful, many non-members from Tagabinet became jealous and wanted to join the group and share from the benefits.

3.4 Outcomes

3.4.1 Accomplishments

In terms of its major accomplishments, TCTA/TURSCO can be credited for the following:

- Significant reduction of destructive resources-based livelihood such as slash and burn farming, collecting of resin from trees, hunting of wild animals and illegal fishing in the Ugong Rock cave and vicinity
- Stopping of illegal fish ponds operations in the Ulugan Bay area within the political boundary of Barangay Tagabinet
- Preservation of the Ugong Rock limestone cave

- Establishment of a community-based sustainable ecotourism in the Ulugan Bay area
- Significantly increasing the monthly income of TCTA/TURSCO members
- Establishing the Ugong Rock Cave Spelunking and Zipline Adventures as a mainstream ecotourism destination for tourists going to the PPUR
- Embedding environmental consciousness and good interpersonal relations to community members
- Creating alternative livelihood for the youth in Barangay Tagabinet

3.4.2 Perceived Socioeconomic and Environmental Impacts of Ugong Rock CBST

In terms of the perceived socioeconomic and environmental impacts of Ugong Rock CBST on Ugong Rock and Barangay Tagabinet the following matrix was constructed based on the information gathered during the focus group discussion. The socioeconomic and environmental impacts were assessed in terms of observable changes within a given 20-year period.

Table 2. Community Perception of Changes in Ugong Rock (Before and After the Project)

	Situation 20 years ago	Situation Today	Explanation for Change
Environment, Agriculture and Forestry	• *Kaingin* of primary growth, upland farming (rice, corn, cassava, banana) • Wild boar hunting • Harvest of *balinsasayaw* nests • Collection of endemic birds (*pikoy, kiyaw,* and *katala*) for trade	• Less occurrence of *kaingin* farming • Lowland farming was encouraged • Planting of high value crops and fruit trees • Livestock and hog raising	• Environmental awareness and Tourism
Fishing	• Subsistence fishing only • Different types of fish and crustaceans are caught (*palos, dalag, ulang, banak*) • Start of fishpond construction • Illegal fishing such as compressor and cyanide fishing	• Tilapia was introduced in the area • Less fish catch • Presence of *baklad* • *Bangus* fishponds are dismantled and banned • Mangrove planting activity	• Effect of fishpond and mangrove cutting which caused environmental degradation
Ugong Rock	• Local park and playing area for the residents • Guano are harvested for gardens	• CBST and tourism site • Sidetrip and alternative route for underground river tourists	• People's organization (TCTA) was organized • NGO intervention • LGU and City Tourism support • Community participation

	Situation 20 years ago	Situation Today	Explanation for Change
Income	• Seasonal and unstable income that is not enough for the whole family	• Employment is provided to members • Monthly income of PhP 8,000.00, with benefits (SSS, PhilHealth) and 13th month pay	• CBST and TCTA
Education	• Only elementary school, no high school in Barangay Tagabinet • Students need to go to the city to continue their studies	• High school is constructed in the barangay • Trainings and seminar are made available to members	• Completion of national highway going to Sitio Sabang, the jump-off point for tourists going to Underground River
Population	• Average household has 7 or more children • Influx migrants from Iloilo, Antique and Leyte	• Average household has 3-4 children	• Family planning

The members likewise listed the following positive effects of the Ugong Rock CBST to members of the TCTA/TURSCO:

- No gender and age discrimination as everyone is working for the good of the organization.
- Spouses are being supportive to members.
- It gave hope to families.
- Environmental awareness is not just to the members but being passed down to member's family as well.
- The whole family of the members are also involved in the organization's activities and the maintenance and improvement of the tourism site.
- Positive outlook among members and their family.
- Personality and skills were improved

3.4.3 Lessons Learned and Best Practices

The matrix below shows the lessons learned from the challenges that faced the TCTA/TURSCO in its more than ten years of existence:

Table 3. Lessons learned

Challenges	Lessons Learned
• Trust among members • Proper communication	• Humility and patience when dealing with other people
• Monetary/Financial Management (Bookkeeping) • Lack of transparency	• Self-discipline and the need to educate self through trainings and seminars
• Dos Palmas kidnapping	• Appreciation to tour guides and drivers for promoting Ugong Rock Adventures to tourists • Realization that alternative livelihood is needed
• Criticisms from others • Interpersonal relationship	• Values formation through inter-faith devotion every Monday • Respect for other religions • Faith in God as a core value of the organization • Practice being a good Christian and it will be reflected in the organization
• Getting Barangay support • Share of the organization to Barangay Tagabinet	• Creation of MOA between barangay and the organization
• Environmental issues (*kaingin*, fishpond, quarrying)	• Environmental awareness because seminars and training
• Zip line Accident	• Safety training for members and putting the guests safety as a priority • Disciplinary action to members who were involved in the accident
• Rude guests	• Customer service improvement • Proper orientation and registration of guests
• Withdrawal of membership	• Series of recruitment for new members is necessary • Determination to reach the goals set by the organization even if only less people are working
• Fear that the government will take over the management of Ugong Rock Adventures	• Need for alternative livelihood and sustainable tourism • Change form association to cooperative is necessary for the organization to survive

Challenges	Lessons Learned
• Death of Dr. Ortega	• Need to be "strong" and independent
• Conflict with landowners on airspace and land occupied by ziplines	• Amicable settlement with landowners

3.4.4 Perceived advantages of TCTA/TURSCO over other CBST projects in Puerto Princesa City

- No need to book in advance, walk-in guests are accepted.
- Transportation is easy because it is near the highway.
- Location is good because it can be an alternative or additional destination for tourists going to Underground River.
- Operation is not dependent to weather or season.
- CBST is marketable because it supports the whole community and identified as an "eco-tourism" site.
- Strong and strategic partnership with other agencies.

3.4.5 Other practices that made the Ugong Rock CBST Project the most successful CBST project in Puerto Princesa City so far based on members' perception:

- The practice of interfaith community sharing and prayers every Monday morning led by a member who is also a Pastor in a Protestant church strengthens the bond and interpersonal relationship among members and defuses stress and tension within the group
- During the time of mentorship of Dr. Ortega he introduced prayers and sharing at breakfast time for the members
- The core values of TCTA/TURSCO were acquired from ELAC, Bantay Kalikasan thru Dr. Ortega and Mr. Tamsi
- The primary values which all members share and to which they attribute the success of the organization are:
  "Values formation through interfaith devotion"
  "Faith in God as the core value of the organization"
- The zip line accident made them more safety conscious and they learned not to blame particular persons for the incident but instead took a collective responsibility

3.4.6 Offshoot Projects from Ugong Rock Cave and Zipline Adventures

The following are offshoot projects that emerged during the operational life of Ugong Rock CBST, all projects are currently operational:

- Sari-sari Store – construction funds were loaned from Bayan Academy and AFI with labor coming from TURSCO

- Zip line Photo Souvenir – construction and equipment funds were loaned from Bayan Academy and AFI with labor coming from TURSCO
- Souvenir Shop – funds were sourced from profits of Ugong Rock CBST
- Fruit Shake Stand – funds sources from TURSCO savings
- Food Service (View deck and Landing Area) – funded from profits of Ugong Rock CBST
- Rice trading – funded by TURSCO
- Micro-lending – funded by TURSCO
- Massage and Spa – funded from profits of Ugong Rock CBST
- Superman zip line – funded from profits of Ugong Rock CBST

3.5 SOCIAL DYNAMICS BETWEEN UGONG ROCK CBST/TURSCO AND EXTERNAL ACTORS

3.5.1 Dynamics between the Community and the Barangay

In the past, Barangay Council endorsements and necessary policies needed by the Ugong Rock CBST were delayed due to bureaucratic red tape. Most members of TCTA/TURSCO blamed the Barangay officials for deliberately delaying actions on their petitions and requests. They perceived that the probable reason was "jealousy" and interpersonal animosities among officials from both the Barangay and the TCTA.

Presently, all transactions between the Barangay and TURSCO have been problem free and expeditious. Members attribute this status to the Memorandum of Agreement (MoA) between them which gave the Barangay 5% of the net profit of Ugong Rock CBST to the Barangay Community Development Fund.

3.5.2 Dynamics between the Community and its Members

During the early stages of the Ugong Rock project, majority of community residents adopted a wait and see attitude towards the project and TCTA members. Criticisms and negative perceptions were common which created tensions between members and non-members which affected interpersonal relationships in the community. Ostracisms in form of mockery and discouragements affected the morale of some members. However, due to the success of the Ugong Rock CBST and the upward socioeconomic mobility gained by TCTA/TURSCO members, community perceptions and attitude changed. At present, those community members who earlier criticized and ostracized the members became members themselves while others became non-member employees for TURSCO.

3.5.3   Dynamics between the Community and Institutional Partners

Unlike in the early period of the Ugong Rock CBST project, many community members observed an apparent lack of support from City Tourism Office. They don't have a reason yet to explain their observations at the moment. There is a close association and cooperation between TURSCO and AFI since AFI's community organizers were deeply involved in organizing and community work since AFI took over ELAC in the partnership role with Ugong Rock CBST. Old partners were: ELAC, PNNI-Pasyar for initial organizing work, Bayan Academy for entrepreneurial development and the City Tourism Office for technical assistance. New and existing partners include the Palawan State University (PSU) and the Legend Hotel.

3.5.4   Legal, organizational, management and technical arrangements of various parties

Legal and technical arrangements between TURSCO and various partners are covered by a Memorandum of Agreement (MoA) either directly or via ABS-CBN Foundation Inc. The following are the existing MoA between parties:

a) **Ugong Rock CBST represented by TURSCO and Barangay Tagabinet**
b) **TURSCO and AFI**
c) **AFI and Puerto Princesa City Government**
d) **The Legend Hotel and AFI**
e) **Palawan State University and AFI/Bayan Academy**

Acknowledgement

The author would like to acknowledge Mr. Marlon Tamsi of AFI Palawan for providing detailed information regarding the history of the Ugong Rock CBST Project and additional information on TCTA/TURSCO. The officers and members of the TURSCO are also acknowledged for providing most of the information both anecdotal and pertinent documents used in this case history documents.

YOUR KNOWLEDGE HAS VALUE

- We will publish your bachelor's and
  master's thesis, essays and papers

- Your own eBook and book -
  sold worldwide in all relevant shops

- Earn money with each sale

Upload your text at www.GRIN.com
and publish for free